Mother Nature, The Interview - Part I

Mother Nature, The Interview - Part I

Oscar Shick

Amelioration Industries Co

Dedication

This book is dedicated to those who will inhabit the planet in future generations.

Speaking their presence will have their existence come forward, and may they be sustained by a renewed planet.

With much love and gratitude

First Printing, 2022

ISBN: 979-8-9870405-0-8

With gratitude to Roger Paul Bailey for the artwork throughout the book and Mona Chavez for the cover design.

Evolution information, merchandise or to submit a question for Mother Nature, go to mothernaturespeaks.org

We have all at one time or another quietly acknowledged Mother Nature, whether it be in wonder of a natural landscape, gratitude for an amazing sunset or bewilderment with the strength and prowess of what we call a 'wild' animal. Our physical bodies are one of the most amazing parts of nature. Unfortunately, more often we curse her for natural disasters, inclement weather, and plagues.

This interview affords her a voice, her voice, and is an opportunity for her to share with the population of the planet. For far too long we have ignored her, poisoned her and denied her cries for help. May this interview educate and remind us.

Q: Welcome and thank you for being with us at this time.

Thank you for the opportunity and for listening. I may repeat myself many times throughout this interview with one statement In particular, I love each and every one of you equally. I don't love you because of anything in particular and at the same time I love particular things about each of you. Each of you needs to follow your path and fulfill what you want your life to be about. I don't believe any of you believe your path is to damage the planet or any of its inhabitants, it may unintentionally happen here and there, but I don't think any of you aspire to have your life be about causing damage to the planet and its other inhabitants. I do think most of you desire to leave the world and all of its pieces in better shape after your departure than it was when you arrived.

I expect some of my statements in this interview may not be terribly popular, please listen to them and understand I may only the messenger in many instances. Life is full of statements and information one would rather not hear nor believe could be true, but that is not possible. With the bad news becoming much more frequent and more bleak, one may believe it is easier to tune it out and believe it does not exist rather than be present to the news. Let me assure you it is not easier nor is it possible to tune out the news on the state of the planet and the environment, the news is not going to go away nor is it going to get any better without some work to make it better. The collective attention span is that of a gnat, in case you don't know, that is very, very short. It is time to prioritize and pay attention to the matters that must be tended to at this time, for the sake of the planet and each of its inhabitants.

Q: *Is that why are you speaking out now?*

Of course it is, take a look around, the planet is a mess! It is a mess at almost every level! The planet has warmed and continues to warm at an alarming rate. The warming is causing many species to have difficulty surviving, it is beginning to impact the human race more than realized yet. The planet works with virtually every thing connected to every thing else. Let me repeat that, every thing on this planet is connected to and hence impacted by every thing else on the planet. Connected may not be the correct word, a better wording may be that virtually every thing impacts every other thing. Nothing on this planet can survive in a vacuum, at least not long term. Something may survive for a short while, but not as long as it would naturally. The prior statement not only makes sense but applies across the board. You can introduce something foreign into an environment and it may live for a period of time, but its chances for long term survival in a static environment is not promising. The human race has been on this planet from its inception and it would be a challenge for it to move to another planet, whether it be a few or many.

Let me go back to the composition of the planet and the interconnectedness of everything. The creation of the energy for the lights in your house may have polluted the water and air fifty miles or more away from where you live, which in turn impacts the health of fish in their ecosystem which in turn impacts birds and so it continues. You may not fully appreciate the impact of your actions until the impact is in your immediate world, until you or someone you know has an immediate impact from the pollution. Driving your car and burning gas will impact the air for many miles. Your polluting of the air impacts the birds, the bees, and eventually the fish in the sea. I am sure a lot of people discount the far-reaching impact of their actions, but I can assure you it is how it works. It would definitely be difficult, and painful, for people to be fully responsible for the impact of their actions and existence. If people were aware of their impact on the planet it would be difficult to process, and troublesome to be responsible for. With the number of inhabitants on the planet, multiplying ones impact times billions makes it challenging for the planet and its systems to maintain.

With the challenges faced by the planet and its ecosystems, maybe you can offer some insights as to why there is such disregard for the planet, and the water, and the soil?

Q: I think there is a plethora of reasons why there is such disregard for the planet and the water and the soil. People don't believe they can make a difference or have an impact on much of anything with all of the frenetic events and activity on the planet. Why do you think there is such disregard?

Too many have lost connection with the planet, its water and soil. It is as though there is no understanding of what you refer to as the order of nature. Too many think they are here to have anything and everything they want, it's truly ludicrous. It is some form of magical thinking and entirely unattached from reality.

There is little understanding vegetables, flowers and trees grow in soil, and water is essential for life. The thinking has become all you need is money and what money can buy, the items it can buy, etc without any consideration for what the impact is on the planet or a person's genuine happiness. If you are not happy, you need to stop and figure out how to get yourself happy. Happy may not always include an ear to ear smile, but often times it will. The happiness I am talking about is the happiness that involves consistent contentment and considerable joy. Joy includes an overflowing of your heart with happiness and joy and the experience of being content. If you are content, everything is present.

If one wants to get reconnected to the planet, one can simply sit and contemplate the circle of life. When someone comes to life, there is perfect combination and alignment of many elements. As they develop and mature, they will become indoctrinated and learned in the customs of society, and often are not aware anything else is possible. If there is belief that nothing else is possible, nothing else may be possible. Once you understand the possibility of a limited or predefined future, you may also understand the possibility of creating a different and potentially unlimited future. It is known as manifesting but it can also be somewhat of a knowing. Knowing you want to or will accomplish certain goals in your lifetime. It can also be a knowing of right versus wrong or good versus bad. Throughout indoctrination, one needs to stay aware and present and think for oneself. Never accept the words of another unless one can feel the truth in the words. You will understand the feeling I am talking about when it happens. It is probably more of a sensation that is really a knowing.

Society needs to unpack the societal collapse of the attainment of money with survival, for themselves. Everyone needs to do that for their own well-being. It starts with an inventory of sorts with what they really need to have their body function, including clean air, clean water, integrous food and shelter. For individuals to contemplate clean air and take responsibility for it is exceptional. It is exceptional and contemporaneously shocking it is not a consideration by most. When life is full of abundance or when ones needs are met without effort or thought, so much is taken for granted by so many. Detachment from the planet has lead people to think anything and everything can be purchased with money and air is one of those things. The same is true for water. When clean air and water are endangered, so is the planet and its inhabitants. There will be little food without clean air and water. As for shelter, it is an unfamiliar word to many. Shelter is a place to shield you from the elements. It has become another opportunity to waste resources and needlessly consume energy. It may be time to have a dialogue on how much shelter any person needs to get perspective on the waste of resources and disconnect between people.

Q: *Are things dire?*

I would not be wanting to be heard if they were not dire and dire may not be strong enough.

My world is upside down and I do not think the potential doom is remotely understood.

There seems to be an inclination by many to only have a couple of considerations when it comes to planet Earth.

The first consideration is whether it is politically correct to support Earth and its environment. It seems to be very politically correct to 'care' about and for the environment. No one can do too much and everyone needs to do something. So given this statement, the question should be 'what have you done for the planet today?'. How great it will be when the planet is a daily conversation and its care ongoing. It is not enough to say you didn't do something that you could have and hence avoided the impact and consequences of the foregone action. If someone told you to thank them because they did not trip you as you walked down the street, you might think they were not one hundred percent. It is the same with the planet and your interactions with it. You should work to minimize your impact on the planet and take actions to help the planet repair itself for your own benefit and the benefit of all. My understanding of politics is it will involve collective actions which are in the best interest of all. There is a sector of the population who act as though there is nothing to be done to help the planet or reduce the impact of society and its existence on the planet. Given the current state of the planet and the fact that the planet needs to be in good condition for the survival of society, it is hard to believe the planet is disregarded and its health discounted. The motive seems to be profit in most cases or denial, but it could also be a lack of understanding or rational thinking.

The other is how difficult it will be given the life one has created for themselves. Very few find any disruption to their life to be acceptable or welcome. Humans have become creatures of habit and creatures of comfort. Putting the same coffee shop, store, gym etc in practically every city is terribly damaging. If every place on the planet was identical, there would be very little adventure or stimulation. If humans continue to deny there are any issues and choose to keep their collective heads in the sand, everyplace on earth may look very much the same though not identical. The unending waste, denial and lack of action do not bode well for the future of the planet.

You have heard the expression 'variety is the spice of life'. Well, a variety of actions will definitely help to keep the spice of life. Actions need to be taken for the betterment of the planet as well as actions to better yourself. Sharing as much love as you can muster will help both. Love, compassion and understanding will help across all areas that need attention.

Again, when you asked me if it is dire, you may want to refresh your understanding and memories of the Dinosaurs, the Mayans and Easter Island to name a few. I don't want to go any further into these at this time, but know that history will repeat itself if something is not done and soon.

Q: So many species are facing new challenges and having difficulty surviving. Many of the difficulties seem avoidable, are they?

Some of the difficulties are directly tied to the hand of humans. Whether it is loss of habitat, pollution or intentional killing, the hands of humans are directly related to the difficulties.

When a situation gets bad enough is usually when a mother is called upon to sort it out. It is when people are ready to listen and be responsible for their part in it all that there can be a change. If they don't want to take responsibility for their part, it often gets worse for them until they are willing to be responsible. I hate to watch it get worse but much is out of my control.

Many on the planet think their lifestyle and corresponding consumption does not have an impact on the planet or the environment and hence there is nothing for them to take responsibility for, such thinking is problematic. Alternatively, there are no guarantees on what the planet can or will provide. Many believe the planet will continue to provide as it always has, as one can see from events in the past years, the planet is changing. Just being alive has an impact on the planet. It begs the question, "why are humans here on the planet?" I will give my beliefs on the answer to that question.

First, let me tell you some of the reasons humans are not here on the planet. You are not here to ravage the planet for your own delight or to pollute the planet for sport. Images of ravaging the planet include trophy hunting amazing animals for no reason at all other than to feed your ego, living in a house that would hold several families or having so many possessions that you have forgotten what all you own. If you have to deprive another living being of life for any reason other than your own survival, you are ravaging the planet. If you are wasting the natural resources of the planet out of a desire to show others you can, you are ravaging the planet. Polluting the planet for sport includes burning fossil fuels for no good reason or just for ambiance, etc.

Q: *So there are questions about where you start and stop given many of the beliefs in religions, can you enlighten us on how it all fits together?*

People spend their entire lives studying and attempting to gain an understanding of how it all works. No offense intended to anyone, but what a colossal waste of time and effort. It would be like reading and studying about how to cook, bake, watching videos and visiting kitchens to really understand it all, but never tasting anything. Understanding is not the same as enjoying or appreciating. Partaking and enjoying and appreciating experiences will bolster ones understanding as well. In some cultures, two people will marry one another, and in many instances, the two will have no knowledge or understanding of the other person or what marriage is. Of course, you can make up anything as you go along but you need to live it, experience it, understand your failures, your successes, and deal with yourself when appropriate. It is in the experience where one learns and grows. So to answer your question, be in the inquiry and discuss and debate with others to make up your mind where it starts and stops.

Q: What are some of your difficulties as Mother Nature?

The name says it all. First, I am a mother, the mother. Children old and young often listen to their mother begrudgingly, especially since they know or suspect she is most often right. The begrudging part often lingers with mothers, often until it is too late. Mothers have an innate understanding of the workings of life and living. Mothers are the source of life. Men have an unspoken belief their sperm is the all precious source of life and it should be treated sacredly. At best, that is flawed thinking. Not only is their part of the formula minimal, but it is also fleeting. Obviously, the mother has the primary responsibility and the burden for the gestation period as well the many years until the maturity and independence.

I will tell you unequivocally to never underestimate or understate the power of a woman. There would be no life without women and they are much more than the provider of life. I should not need to state this but the continuing poor treatment of so many women demands I speak up. Women are equal in every way to men, mentally, physically etc etc. The subtle and not to subtle ways men treat women to demean them or convince them they are less will come to an end. Men need to treat women as equals and women need to believe in themselves and act as equals. The treatment of women as less than has gone on for far too long, the species cannot continue with such treatment. I love you all and you all need to love one another.

Q: *What would you like to have understood about the planet?*

Virtually everything on the planet is living, yes everything. Living just as every human is living. Every thing is on its own journey, on its own timeline and with its own quest.

There is a notion that in order to be living, a mouth, lungs, etc must be present. The presumption is it needs to be just like you. In making these statements, it becomes clear it is not true. A tree is an easily understood example of a living being that has no mouth, lungs, etc. I doubt anyone would deny that a tree is living. I bet many would deny the soil is alive, but it is very much so, just as all things are.

Since everything on the planet, including the planet itself, is alive, there needs to be cooperation and respect amongst all. The prevailing attitude of most of you that the planet and all on it is here to serve you and for your pleasure must not and can not continue. If you want the planet to take care of you, you need to take care of the planet. It is simple.

Q: What would you like to have us understood about our-selves?

Each of you, every human, is on a journey. The soul is furthering the journey while here on the planet. There is not a lot to disclose about the journey, except suffice it to say you should strive to make it as full of learning and growing as you can. As to learning, it is not all book learning though there is much to be garnered from book learning. The other learning I would encourage is what might be referred to as life learning. Learning about life, what life is about, what is possible for your life, expanding your heart both by loving others and letting others love you to name a few. Book learning can happen entirely in solitude. While you can read a book and take away from it whatever you understand, your understanding would probably be deepened or widened in discussion with someone else who is also learning or studying the same subject. Life learning must absolutely take place with others. Out of your response to others, you will learn about yourself. Occasionally stop and critique yourself, see what you did perfectly and what could use some improvement. I am sure you have had much different responses to the same words from different people.

Q: Anything else we should understand about ourselves and our relationship with the planet?

There is a typical life cycle for humans that is well documented just as there is a life cycle for most any living being on the planet. There is also a life cycle for the planet, a sobering thought.

I am sure there are countless questions to arise from those statements. So much debate is possible, but, it is not needed. If humans could consider a life where there is respect for everything on the planet, including every other human, every plant, every other animal and so on and so on, imagine what would be possible. Respect in that every human, plant and animal has the same right to survive and thrive on the planet just as every other human, plant and animal. It is really quite simple. There are an almost unlimited number of varieties of plants and flowers and the planet is better with the variety rather than the monotony of the same varieties. It is the same with humans. The planet is better with all of the varieties of humans on the planet as opposed to one single or very limited variety. There is always something to be learned from another whether it be through observation or interaction. Always be kind all ways.

As respect and eventually love replace disregard and rudeness, the energy it will create will be virtually unstoppable. The energy of war and conflict could actually be overcome. There is no need for war and the destruction that comes with it. I doubt anyone derives great joy from seeing destruction and suffering. Few think about or realize the impact of wars on the planet. The number of years it takes for the planet to recovery from war is shocking and a needless setback to the planet. I can almost sense the mockery and disbelief this will be met with but I cannot plead strong enough for it to be attempted. You were built with love and respect but somewhere along the line it got muddled or morphed, return yourself to your roots.

*Q:	Are there other places in the universe to go and live?*

Of course, although rare and not equal to Earth. There seems to be a belief there are other places to go and humans can pick up and go to another planet when done here with Earth and have the same quality and richness of life. Life here on Earth is easy, there is air, water, vegetation grows easily and bountiful. There is also beauty in abundance.

Other planets, including those potentially suitable for human habitation will involve a constant struggle to survive. A struggle for food and shelter and a lack of beauty. I understand the phrase 'beauty is in the eye of the beholder' but it is difficult to match what has been here and is holding on. What one may find beautiful another may not. One may not find something beautiful for many reasons, including a lack of context, a lack of understanding or a lack of appreciation. It may be time to reacquaint oneself with beauty in a way that you may not have in your past, whether it be at a microscopic or a vast level.

Q: How about a few silly questions. Have beings from other parts of the universe come to Earth?

I think you and everyone else knows the answer to this question. There is more on this subject but I think it best I wait to go into it further until a later date. The question on beings from other parts of the universe or aliens you may want to ponder is what hospitality would you show them. A being coming from another part of the universe may be far more complex given they have the means to travel to Earth.

Q: What about weeds? Are they an intentional part of the landscape?

Let's start by looking at the definition of a weed. A weed is a plant considered undesirable in a particular situation, or a plant in the wrong place. Who gets to say what a weed is? When I came up with many of the plants, I didn't waste my time and just create nuisance plants and other undesirables. There is a complex and complete balance to it all. Rather than try to have the world look the way you want it to look, strive to understand the world or even just accept the world for all it is. Also accept the world for all it is not (for you). Ask anyone who has been in a relationship with another for a period of time, I would bet they would agree it is a lot easier, and more joyful, to accept their significant other for who they are rather than try and mold them into who you may want them to become. If you are in the right relationship, that will be a natural perspective. To answer your question, everything is an intentional part of the landscape, and you get to choose whether to call it a weed or not.

Q: What about pests?

With the same thinking as weeds, a pest is defined as a destructive insect or other animal that attacks crops, food, livestock, etc. It could also be an annoying person or thing. Even though pests are pests, they serve a purpose in the order of life be it food for others or food for thought. There is often a silver lining or gift to be had with many pests in your life, don't miss it.

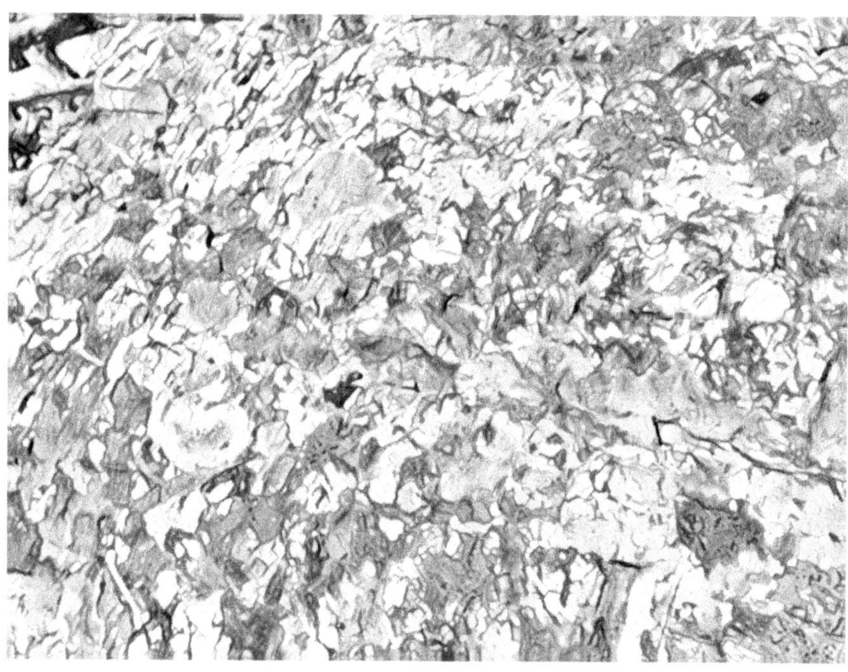

Q: *Are viruses part of nature?*

I see we are back to a more serious topic. Of course, viruses are a part of nature. They consider themselves a sort of protector of the planet and especially me. When the planet gets out of control or better said out of balance, a virus often steps in either by design or by consequence. Historically when one species has had a disproportionate impact on the planet, they become their own worst enemy. Humans could find themselves in a predicament in short order. The needless and non-stop burning of fossil fuels has an impact on the planet that is hard to convey. For example, if someone is in pain and their body has grown accustomed to the pain, they probably will not understand how much pain they are in until the pain is gone. In similar fashion, the planet has become more and more polluted to the point that many people are not present to how polluted the planet is as they have become accustomed to the pollution. People don't think there is anything they can do to help or think it is hopeless. The only thing that is hopeless is people thinking they are off the hook if they are hopeless. Being hopeless is not genuine, it is a way people try to let themselves off the hook for being irresponsible about something. Individual responsibility is one of the traits that set humans apart from others species on Earth, especially in making choices as opposed to survival instincts. Many share responsibility with others or have a group or collective responsibility, even with shared responsibility you need to take full responsibility.

Q: *Why is COVID happening?*

The why of it all may be an oversimplistic question. What is the why of anything for anyone? Every being on the planet is on a journey and hopes to learn lessons whether they are present to them or not. So to respond to a question with a question, what lessons do you believe there are to be garnered from COVID? I am not referring to what lessons there may be in the immediate response to COVID or anything along the lines of immediate responses but more in the realm of what there is for society to learn by its response or what individuals can learn about themselves or others.

When an individual learns something about themselves, it is not in a vacuum. Learning about oneself does not occur without help from others. It is impossible to be alone, although many believe in such a façade. When you learn something, discover something, about yourself or others, there is the context within which it occurs which makes it impossible to learn about oneself without others. The comparisons and contrasts to others forms the basis for the discovery. There has been plenty of time for you to compare and contrast with others and make some discoveries about yourself.

Wanting to leave your house and 'get on with your life again' is amusing. As brutal as it may sound, most of you have been wasting your life with unworthy pursuits and misguided desires.

You may want to consider some of the following as potential COVID-19 lessons.

A change may be necessary.

A change may have occurred and this is part of the impact. The changing climate and its impact on the planet may wreck havoc on protections or systems of the planet. Viruses that may have been frozen and locked up for many millennia may have been or will be released as a result of warmer temperatures and melting. Plants etc may have migrated in order to survive and their new environment provides an environment ripe for mutation. As one changes, many others will follow.

A breather for the planet was necessary. The air actually cleared when many were home and not in their vehicles wanting or needing to get somewhere or nowhere. Just as your body rests at night when you are sleeping, the planet also needs time to rest and catch its collective breathe to thrive another day. The inactivity which accompanied covid provided the planet some rest and the results were promising even if short lived.

The planet may have reached its capacity, or be over capacity. So few believe there is a capacity, and there are a couple of ways to look at capacity. There is little or no capacity if there is no concern for space to live or cleanliness or food supply. When there is a desire to have more than a closet to live in, have sufficient clean water to drink and bath in, and have food to eat, there will be a capacity limit. Clearly the planet has a capacity limitation and it has been exceeded. In part due to the size of the planet and also in part due to the unmitigated pollution created by the population.

Finally, have an appreciation for the world you live in, as all lives are precious, oh and that it's just as easy to be nice to one another as it is not to be.

Pandemics and viruses can be blunt and direct with their messaging and impact. Your question appears to be rational, but is really beyond the comprehension of many. A rational answer to an irrational question or topic can often be irrational.

Q: Do pandemics happen for a reason?

Of course, pandemics happen for a reason. The reason can vary from person to person, culture to culture and species to species. Much of the variance can depend on the degree of awareness present and who is present and to what. Everyone is in this together whether it be a pandemic or daily life on the planet.

Assuming at some point growing up, you looked at a goldfish bowl. In that bowl, there was not a left side and a right side and they were not independent of one another. The entire bowl was one, one environment for the fish to be in. There is a sense of oneness and you should ask yourself if you are present to it. I am not only talking about the fish tank, I am talking about your life and your world.

Humans have a way of coming together in times of difficulty or disaster, but not being able to fully swing it otherwise. It is something each and every one of you needs to work on. The discourse is an old energy and thought pattern that could easily leave the planet if it was not continuously generated and consistently honored.

Of course, there are differences of opinion on the subject of oneness, but that is different than a realization that everyone is in this together. I don't think any of you would get on a plane and think that the people in the rows in front of or behind you could parish and you survive the plane falling out of the sky. Such a thought is not rational and it also parallel to all of you being on the planet. It is not rational to think that some of you can avoid the impact of changes on the planet while others will bear the full impact. It is not only irrational but also impossible.

Q: *What will fill the void left behind when discourse is no longer?*

One word, love. Oh, I can feel people roll their eyes. Love is much more complex than people think it is and simultaneously much more simplistic.

Q: How is love more complex?

Love is an all encompassing yet pervasive energy. Love is part of most all feelings and emotions. One can love the way someone makes you feel, whether you are in physical proximity to them or not. One can love to hear from another and simultaneously love to get together with them or love that you are not getting together with them. Love underlies most everything. It is complex in its pervasive nature.

Q: How is love more simple?

Love is all encompassing. It is everywhere all the time. Even when one loses someone most people unknowingly struggle with the loss of love even when it was not a romantic relationship. Love is why you are here and what you should work to be present to at all times. When presented with love, many ask if that is how they want to be loved. There is probably much more love present than you allow yourself to be present to. Even when someone is upset with you, they may be standing on a virtual pedestal of love. They may be committed to something else for you in the name of love and upset you do not have it or choose not to have it. It can be awkward. The same awkwardness can be present when expressing love to another.

So many spend their entire lives before they understand love when it is really quiet simple. Yearn to be present to love on a moment to moment basis. Love is all around you at every moment, start with not pushing it away or feeling as though you need to ignore or quash it. When someone tells you they love you, let yourself feel it. When you tell someone else you love them, look at them like you mean it and you want them to really know that you mean it.

Q: It can be really difficult to love someone all of the time, how do we do it?

I can imagine it is difficult, but you should strive to love everything all of the time. Love is not the same as like. Love is not the same as accept, nor is it the same as acceptance. Even when someone is selfish, destructive or mean, love is underlying it, although few are present to the love that is present or their view of love is misguided or misunderstood at that moment. When one is selfish and taking more than they need, they may do so because they think they love what they are taking. Love can be a matter of taking a stand for someone who may not be present to love at that moment and showing them love and compassion. I am sure you have heard others say it, but love wins over all else.

Q: What else goes with or is often present with love?

Forgiveness can be a big one that is present with love. It is part of the journey, part of the lessons here to explore and learn about. You will learn from those who are unkind or even downright mean to you. There is often a lesson in your interaction with them whether it be something you learn about yourself or something you revisit from a prior interaction in your life. It is a lesson. It is not a burden you need to carry with you for the rest of your life. Learn your lesson so you become a better person to yourself and others. This is an example of some of the energies that need to dissipate from the planet. Carrying negative thoughts or resentment against someone will impact you in an adverse manner, it will not impact the person you hold in contempt.

Respect is also often present with love. Respect starts with a basic honoring or respect for life and the planet. You should include yourself in this and respect and honor yourself and your body. Treat yourself in the same manner in which you treat others and in doing so elevate how you treat others. Do not find yourself swayed or influenced by those who do not treat themselves or others well, effectively disrespecting everything. During covid, many in charge did not take precautions instead letting people die unnaturally. The lack of respect for life was heartbreaking.

Beauty is also often present with love. There is beauty in so many things and so many places. Beauty is often one of the foundations or justifications for love. Something is so beautiful that you love it. I want to make one thing clear here, everything and every place is beautiful and full of beauty, it may not always be obvious at first glance, but trust me it is. When you find beauty in everything and appreciate all for what they are, you indeed are loving.

Q: What do you find to be most beautiful?

Beauty is not a competition, again there is beauty in everything from the prismatic colored flowers that bloom to the pad on the foot of a tiger or cheetah. Many will only see beauty in the flower, the grass, the trees while others will see the beauty in the worm, the soil or the moss on the rock. There are no limits on beauty. Beauty is in everything and everyone. I am not stating that lightly. If for some reason you don't find the beauty, look again, possibly change your perspective, squint your eyes, tilt your head, stand on one foot and you will find it. Hopefully you need to do no more than just be present and allow your senses to be aware of all of the beauty around you.

Q: What more can you say about the pollution?

I will make this very simple.

The sky should be blue and vast.
Water should be clear and pure.
The soil should be full of life, not toxins.
Vegetation should be free to grow to its natural confines.
Mammals should be thriving and fulfilling their intent to do what they need to do to survive and live another day.
Sea life should be abundant and free to roam the waters without being surrounded by poisons and toxins.
At the risk of sounding my age and old fashioned, some things should not change.

You know it is the way it should be, but it is not. The reason it is not is in large part due to the impact you have had on the environment. You need to be realistic and responsible for that impact and understand that your impact on the environment and the planet impacts every other living thing on the planet. I know you have heard that before and I am sure it was far too ethereal for you to understand. What does your body tell you when it hears these words? Your body will not lie to you and your body knows what is best for it. Your body is probably the only one who will not lie to you or spare your feelings. For this reason if no other, you should consider being kinder to your body. For those of you who just giggled a little thinking that is a ridiculous sentiment, you are the ones I am speaking to the loudest.

Q: How do you define pollution?

Initially I find myself upset with this question, it seems to be too obvious to me and at first glance unneeded. On second thought, an understanding of pollution or lack thereof may very well contribute to the problem. If pollution is justified or excused because others also pollute, it creates a race to the bottom. Pollution should not beget more pollution or excuse pollution on a smaller scale. I believe pollution to be when something is no longer in its natural state and not creating a symbiotic relationship with something else.

Q: What is the cost of acting versus not acting?

Might I suggest you rephrase your question. When you include the word 'cost' in any question, there is an implication that it can be reduced to dollars and cents and that the only cost is a matter of dollars and cents. Again, there is a misguided belief that the world comes down to a matter of money. It does not.

There is a cost to a storm, a flood, a fire or whatever you refer to as a natural disaster. That cost can be partially computed in dollars and cents. Only partially because the clean up is what can be equated in dollars and cents. What is the cost of a fallen tree? a reshaped canyon wall? a lost habitat? Only a simple or deceitful mind would think you can reduce any of that to merely dollars and cents. If you look at it all as a matter of dollars and cents, you are already bankrupt as there is not enough money to restore the damage caused to date. One could also say you are morally bankrupt and soulless.

If there are no actions, there will be a future unlike any you can imagine. You may have seen someone's interpretation of what the future could be, but it will be much worse than anything you can imagine. I hesitate to say those words in as much as many of you not only won't believe what I am saying, but you will also taunt it to happen and see how bad it will have to get in order for you to understand it will be bad. To those of you who act in such a manner I say grow up and open your eyes to your behavior, mostly bad behavior, and get responsible.

Consider living within a context that the planet is sacred and every living being, including you, matters as they live today to reach tomorrow.

Q: What we have come to know as COVID-19 has wrecked havoc on life as we know it currently, what can you tell me about it?

Since the early days of the planet, viruses have been abundant and very active on the planet. Species developed as a result of viruses, directly or indirectly.

It is also another little known or unknown consequence of a rapidly changing planet and its ecosystems. Viruses are living beings, just as humans are living beings. Humans are not the only living beings that will do what needs to be done to survive. Viruses can ever so patiently lay dormant for centuries or millennia waiting for an environment suitable for them to flourish. Once conditions are ripe they return and will adapt to their environment so they may flourish. While humans have the ability to do so, it is often perceived as 'too expensive' and not undertaken in a manner that will have human beings succeed. My definition of success is not winning one season, one day or one race. My definition of success is maintaining a planet to allow all to not only survive, but also to thrive. The experience or memory of a biodiverse planet should not become something that will only be known in a museum.

The planet is living as is everything that makes up the planet and lives on the planet, a big ball of life.

Q: There is currently somewhat of a fascination with exploring other planets with the possibility of having humans move there and start a new civilization, thoughts?

It is simultaneously hilarious and saddening. It is hilarious in that it is relatively intelligent people who propose exploring other planets with the possibility of having humans populate another planet. I think most anyone who is intelligent has moments where their book smarts do not shine through. If a planet as perfect as earth was located and was feasible for habitation, such a minute number of persons would be able to relocate there it would be dismal. Also, there would be no diversity of life.

The thought of humans habitating another planet is saddening, knowing it would also get screwed up. Many times someone will fail at something and then repeat the exact same actions hoping they will not fail again. I am not here to ask the questions, but sometimes posing a hypothetical question is the easiest way to have another understand, even if the question goes without an answer. Not only will any move not go well for humans, but it will also accelerate the decline of conditions here on the planet. The powers that be will also screw it up in large part because they will try and make it something it is not. There is no place that can be made to be this planet with all of its attributes. Any other location will solely be a place to survive, not a place to thrive.

I have always taken care of each and every one of you, even if it was only to provide you air and water. Have also provided the building blocks for life and its many accoutrements.

Humans seem to have difficulty accepting what is as well as accepting what is not.

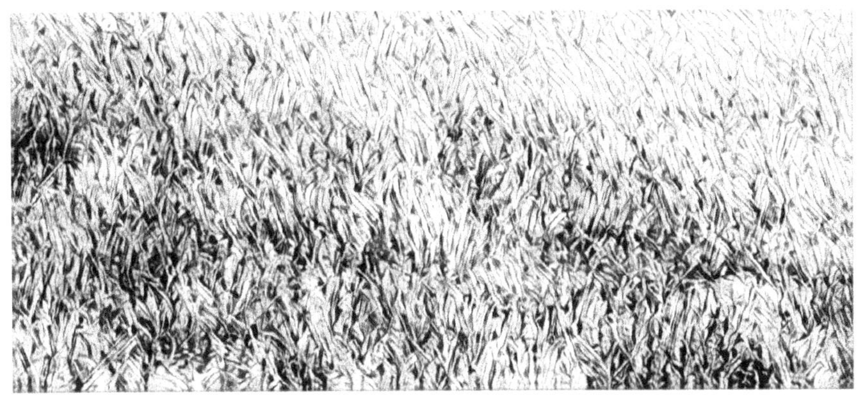

Q: The word present has come up several times, what are you present to on the earth?

Let us presume there is nothing on this planet that was not intentional, you may also use one of the words designed, engineered or created in place of intentional. Let us be intentional and let us start with being intentional about being present. There is no need to be more than present at this time, practice it, observe it, delight in it. Being present starts with being aware of possibly a minute detail or the biggest detail. Be present to nature, start with studying at the structure of a leaf or savoring a sunrise or a sunset. Whatever you choose to be present to, be with it as though there is nothing else in the world for you at that moment.

Most here have forgotten why they are here. There is no one who was placed on this planet to make and accumulate as much money as they can. When someone accumulates so much more money than needed, they are sending messages to others. The messages are not positive, though many believe they are.

An unintended consequence and impact of the accumulation and concentration of wealth by anyone causes others to suffer a deficiency or a lacking. The accumulation is a game without a winner. Those who accumulate may believe they have the experience of a win, but they only try to convince themselves of a win. The win, for lack of a better word, is merely an acquisition of a physical item.

Q: Is there hope or despair on the planet now?

There is both, there is always both. One will be stronger than the other and then vice versa. When people have hope, they carry on believing they will get somewhere to someone. Hope can be the sustenance to carry on yet despair can also sustain one to keep going. Whichever is present, there is most always love present. Love gives one hope to keep going just as despair will. All want to love and be loved and will go to great lengths to do so.

Q: Many are having a difficult time with their mental health right now, any recommendations to help?

It needs to be understood expectations are very much the source of discomfort and dis ease that so many experience. Expectations that life will go a certain way is almost certainly foolish. I don't believe one should have detailed ideas or plans on what life should look like. So many of the best moments are those which could not be planned. So many try to plan falling in love, the ideal trip or the ideal life and, for them, all is measured against these ideals. It is not a path one should strive to get on or stay on. So much can be missed when strictly sticking to a plan developed without experience.

Q: What about extinctions, are they natural?

Extinctions are as much about life as they are about death.

Think of extinctions as part of a huge circle of life. The dinosaurs went extinct as a result of a virus more than as a result of a small meteor hitting the planet. I am sure it is difficult for you to believe a virus could be such a threat, but they are more powerful than you can fathom. To be clear, an asteroid hitting the planet can be a very serious event for those living on the planet and it makes a huge mess. It certainly has an immense impact on the land, water and the air. A virus can also have immense impact on a planet without a lot of coordination to minimize its effect.

Q: Is the planet overpopulated?

Very few are upset or take action when many are sickened by the pollution in the air.

Very few are upset or take action when many do not have water available to drink.

Both air and water are necessary for life.

The number of inhabitants on the planet is taxing the planet and its inhabitants. It raises the question would you continue to do something if you knew it was hurting or even slowly killing something, someone, anything? I am going to break the news to you that you do.

I don't care how much someone may represent otherwise, ordering everything you need from a website, having it wrapped in plastic or mountains of cardboard has a monumental impact on the planet. Yes, you are not getting in your car and driving many places and polluting the air. There is a vehicle coming to you to deliver one item, in a box, with tape on it, a printed label, etc etc.

If you take offense to what I just said, part of me is smirking and part of me is grimacing that you are missing the message. Instead of playing the victim and lamenting how much your life may be impacted, celebrate and welcome your voice in a new chapter. It is time to get real about what is happening on and with the planet and participate in its nurturing.

So many are missing the messages as the climate changes and instead wanting to be the victim of something. There are many victims of climate change, but not as many as claim to be. Being concerned you may not have your favorite coffee on your drive to work in the future does not make a victim. Being concerned coffee fields where your coffee comes from may not produce in the future as a result of climate change may make climate victims of the farmers etc who bring the coffee to market. If you were to count everyone in the production of the coffee, the impact will be surprising and vast, yet you will find another satisfying morning beverage. This is another example of the connectedness and resulting oneness of everyone on the planet. Imagine the impact of climate change on those who work outdoors who will endure higher or lower temperatures, or on those who cannot afford to pay for additional cooling or heating, or on those who will have little to no food as a result of fewer crop lands. The impacts are real and heartbreaking and will get worse unless humans come together and work with one another to mitigate the impacts and restore the planet sooner than later. Unfortunately, time is running out.

Q: ***What messages are there with climate change?***

No matter what name it is given, or what it is called, it is real and it is happening. I prefer to call it by some of the other terms or phrases that are more descriptive of what it is, 'planetary demolition' or 'extinction taunting'. I know both of those are a bit in your face, but only after it is acknowledged can remedial steps happen. I doubt these are messages other than to tell people to wake up and realize what is happening before it is entirely too late.

If you do not do anything about it or change the way you perceive the changes, nothing remedial will occur, nothing will get better or even stay the same. The ball is rolling down the hill and at some point it will hit a rock and bounce out of control.

The planet is changing and it is largely the result of the actions and inactions of humans. You are the only species on the planet that can cause the changes that are occurring and the only species that can do something about the changes that are occurring. All the other species on the planet have not changed their impact on the planet and most have reduced their impact as a result of their numbers being reduced by humans. Many have been run into extinction as a direct consequence of the barbaric actions of humans or inconsiderate encroachment onto their habitat, but they are not burning fossil fuels or taking actions that harm the planet.

I have heard many past generations state when they get older that there will be enough for them while they are here on the planet and hence they did not need to concern themselves with any behavior modification. A time has come where that statement may not be true. There may not be enough for you while you are here, whatever resource that may be and whatever age you may be.

It is time for you to be personally responsible for your time here on the planet and the impact it has on the planet. No one needs to make you responsible for yourself and your actions, you need to do that for yourself, now.

It is also time for you to foster and understand your role in the collective responsibility, also called social fabric. In understanding your role in the collective responsibility, you will come to understand the impact you have on others. Whether you encourage others to do the right thing when it comes to daily life, the environment and upholding the understanding of the complexity of the interrelatedness on the planet.

May very well be more to life than what you think there is. I have watched so many for so long be concerned about so much for so few, or is it so little concern for so many. I see there is a lot happening on the planet and it is not only difficult to keep up with all of it, but it is also difficult to process or understand it all. When the choice is long term versus short term, it seems like short term wins most every time. When the choice is plan for tomorrow or survive today, it seems survival wins most every time. It is becoming evident water will be an issue for many in the very near future. Instead of building pipelines for fossil fuels that are destructive to the planet, it may be time to consider building pipe-lines to move water from locations with an abundance to areas with critically low water resources. It is hard to believe such pipelines have not been built, it is not hard to believe no one will want to build such pipelines, at least not until they are thirsty.

Q: I know we are all going to die at some point, is that what you are referring to?

Partly and yes, everything goes through a life cycle and that certainly includes death. It is certainly possible, and probable if the current trajectory continues and the human race does not start paying atten-tion, much of the human population may not survive the changes of the planet. That is beyond my comprehension but if no one starts to remediate the planet and in turn save yourselves, the writing is on the wall. It is on the wall with bold metallic letters yet few can or want to read the text.

Q: WOW and I am shocked. I don't think many believe the state of the planet is that dire. Is it really so dire?

I know it sounds dramatic, but humans have been slow to evolve, in large part because they believe they can outsmart everything and still maintain their standard of living. Humans actually have no other choice except to evolve. I think that came out in a confusing way. I am in no way saying human beings will not continue to occupy the planet. What I am saying is the future for human beings on the planet will not continue to look the way it does today, it can't. There has been far too much damage and destruction to the planet. If there were a significant commitment to stop the ongoing damage and work to reverse it, the planet could come back to being close to what it once was, but there is little indication of any coming together to accomplish some loving for the planet.

That sounds fairly discouraging but I think most would agree that there are too few working together to make the changes that have to occur in order for the planet to get better.

Q: If people don't come together to make significant changes, is there any point in people working to make betterment on their own?

I don't want to discourage anyone from taking steps to help the planet because every single effort will help, no matter how small. Just as every adverse action has an adverse and cumulative impact, every action to remediate has a positive impact. Without being negative or pessimistic, the planet is in real trouble and it will take a lot to reverse the decline. Just like with your body, it takes something to come back from an injury or illness, and that includes consistent efforts, mindfulness and intention.

Q: Can you expound on those three?

Consistent efforts is doing something consistently or in the same manner over and over again. Possible examples that would help the planet would be walking more to get where you need to get, planning trips so you don't get in your car or other transport to go get a few items here and one there. I am sure you have heard all of the many ways to save energy etc. Saving energy is big for the planet and so much carbon is expended to make the energy which is often wasted. In addition to being aware of your energy usage and the impact of wasting it on the planet, please consider being mindful and intentional.

When I refer to mindfulness I am referring to being present. Being present can be difficult for many to grasp and my incorporating it into being mindful probably does not help. Being present includes being aware, aware of your surroundings, your place in your surroundings and others in your surroundings. Being present includes connecting with your heart and including your heart in your interactions and daily life. Simple actions to be present include:

Find balance in your life and living.

Show kindness to the planet, your own self and others.

Cultivate respect for yourself and the planet.

Too many of you wait until your final days to acknowledge yourself and your personal actions and history. In your early years your eyes are wide open and well as your mind and your heart. If any of them have closed somewhere along the way, you should work to reopen, and you can do so with simple intention. There is no need to go on a hunt to see when it closed and why, or who is to blame, just set your intention to reopen, and enjoy your life. You can set your intention in most any area of your life.

Q: *Do you follow any of the politics in the world?*

I don't follow any specific politics per se, instead I find myself watching the people and their well-being. The well-being of most all in the world is impacted by politics. I do not understand the unkindness of it all. Just as you need to get along with nature to co-exist, you need to get along with each other. You don't need to be besties with each and every person you encounter, but you should respect their life, their beliefs and their presence. It is very basic.

When a person does not respect another person's life, in fact, any life, they are not allowing the world to respect them back. Some people tout 'you get what you give' and other similar phrases. However someone wants to phrase it, it is one of the ways your time here on the planet works. You can test this by smiling at another, greeting a stranger with a wave or a good morning. Give a dog or a cat a good rub and see if you have a new friend. Allow yourself to be present to the smile growing on their face or the general lift they may show just by being acknowledged and respected.

I haven't answered your question, I am just really excited to be heard. Back to politics, it seems like politics have become very primal. It has become a game of win or lose, a sort of primal survival, as opposed to governing and fostering an environment for people and the planet to grow and flourish. I may repeat myself, but people will not flourish for long without taking care of and having the planet flourish. Governments should make sure the planet is taken care of and nurtured so its inhabitants are able to flourish. It is not a radical idea to take care and nurture ones habitat. It will not be long until governments will come to understand the limits of the planet and the impact on humans when the planet adapts in order to survive. Many nourish the planet but many governments simultaneously attack and damage the planet in countless ways. Growth, whether it be in population or widgets produced, is mostly destructive or the precursor to demise. Like the child who discovers sweets or ice cream for the first time. A taste for their initial introduction would be advised as a larger serving will tax their system and make them sick. It is the same with the planet. The planet has limits, it is finite. I do not understand why this seems to be a concept so difficult to understand by so many. Even the planet, as vast and massive as it is, has limits.

The limits highlight the need for rational yet heart centered decision making. Many leaders make their decisions based on what they believe to be best for themselves and their desire to continue to occupy their elected office. Such decision making is that of a politician, not a leader, civil servant or necessarily an elected official. I regret to inform you a politician rarely has your best interests in mind and never has mine in mind. A politician will line his or her pockets and work to get themselves reelected so they can continue to line their pockets from the riches of few rather than do what is best for the many. You should work to not have any politicians represent you. Alternatively, a leader will lead, hopefully in the best interest of the community and society. I say hopefully as nothing is guaranteed but there is a bar against which to judge their actions. The bar involves identifying whether they are making choices and making a direction for society which includes heart and care of the planet.

Heart does not necessarily mean there is not strength, to the contrary, heart is all strength. Every one of you needs a heart to survive and since it is a significant part of your being, it should have some say. Many of you probably feel as though you followed your heart when you were younger and it may have led you astray. Your heart did not lead you astray, your minds interpretation of your heart lead you astray. It may have sabotaged your heart in order to remain in control and lead you where it wants you to go. Listen to your heart again and follow where it may lead you.

Whew, I know this has been a long answer, thanks for bearing with me.

Q: So you mentioned decisions to be made with the heart or with heart energy, what more can you say about that?

First, let me clarify, there are not only certain random matters that you should allow your heart to steer you. Your heart should really be a part of decisions you make in most every area of your life. Many believe your heart should steer you when it comes to matters of the heart or love or charity, period. There is probably only one way you should not exercise your heart and its energy, you should not be stingy with it. I am sure you have heard that love conquers all. Add me to the list of those stating that as the truth. Most every decision and choice you engage in should include heart, and hence love. I know it can be a more difficult process, but you know it is better for the planet and it makes you a better person as well.

Making decisions with heart includes considerations of both your heart and the heart or hearts of those the decision will impact.

Back to politics and decision making, and as I love everyone equally, it would be great to see governments be more representative of populations, hence fewer old white guys.

Q: Excuse me, what do you mean fewer old white guys?

Of course not every old white guy is the issue, but it seems like most want to hold on to whatever perceived power or authority they hold. I think I have stated before the old white guys have royally screwed it up since they have been in charge and making the decisions for so many centuries. Most of the old white guys are more concerned with their ego and accumulation of money than they are with the welfare and happiness of the population, including those who do not look like them. I am sure it is plenty scary for them to come to understand they may not be in control forever and they may actually see the end of their control in sight.

The world is plenty fluid and changing as well. It is time to entertain new ideas, re-evaluate most everything and forge a path forward together. Most in the system would rather win than figure out how to take care of everyone. Far too many have been invited to the dinner party and not been served more than a soup starter.

Q: It sounds like a radical idea to purport that those in power should wrap it up and move on, knowing that what they may be wrapping up may be undone in short order. What makes you think that is an idea whose time has come?

As I stated early on in this interview, things are a mess. It is not practical or rational to think the same people doing the same things will bear a different result. Those who have been in power have not taken care to plan for and anticipate the future. There is a dramatic difference between someone planning for their next ten years on the planet versus their next thirty or hundred years on the planet. Individuals can plan for their own time yet they also need to consciously plan for and include the planet and its future.

Many cultures have been and continue to be short sighted and incapable of taking care of the people that live within their borders. The culprit in these cultures is their leaders. If the leaders will not lead, they need to be replaced. It is simple, if the job is not getting done, it is time to get someone else to do the job.

The short sightedness and inability of leaders to anticipate future needs and the consequences of failing to act can be horribly damaging to a society. If a society and its citizens are only concerned about today, it may not survive to see tomorrow.

One part of the rationale for leaders needing to change their thinking or have others take over is because nothing is forever. There is a belief that the planet in its current state will be forever and that belief is being challenged. Between the pollution and the continued burn of fossil fuels, the planet is destined to change at a pace that is unfathomable. The part few understand so far is the impact the changing planet will have on humans and other living beings on the planet. I think I stated earlier how everything is connected and an impact on any one item will impact most if not all the others. I strongly encourage you to give some thought on what could disappear. Specific items, such as plant and animal species will undoubtedly disappear initially. I bet most would find that to be an acceptable loss. Those initial losses will be followed by more losses, so on and so forth. The thought process needs to go from what may disappear to how will life be impacted. You will be surprised to see what will disappear from your life in terms of products and experiences and the impact on the quality of your life.

Q: So what will help to mitigate the losses and the collapse of the planet?

There are the obvious things be done, saving energy, burn less fossil fuel, plant trees, etc. All of those are important and should be done consistently and often.

The less obvious yet more important items include compassion and empathy. I want to remind you again that you are all in this together. There is not a single one of you, no matter how much money you have, that can insulate their world and their life from the impacts of the changing planet. The impacts will touch each and every one of you, although probably not equally which brings me back to compassion and empathy.

Each and every one of you can have a profound impact on your world and those in it. You can understand the impact your actions and in many cases your selfishness has on others as well as the planet. Once you understand the impact of your actions, you will see there is another way and you will not need to continue to be a part of the problem. You will also see how connected you are to others on the planet. The connection may sound like a burden, I assure you it is not. When you work to expand your compassion for others and find empathy as well, your life will expand in a way you may not be prepared to discover.

Q: A dear friend of mine started a conversation with me when I was in my late teens that I presence myself to often, that conversation was "why are we here?" So why are we here on the planet?

At the risk of sounding like too much of a hippy mama, you are here to have what I have heard referred to as a human experience (if you are a human). While you are here you may have many choices, as such you do not come with everything set in stone. Your experience or lessons may be set or predetermined, but not how the lesson looks or how it unfolds. Much like going to a party, you hope to have a good time, but a good time may look very different from one party to another or from one attendee to another. Some parties may have you as the star and others have you as the audience. You only need to be present and see what is presented to you, at parties and in life.

Would you watch a movie if you knew how it ended? I am referring to the first time watching the movie as opposed to watching the movie the second, third or fourth time for one of many reasons. Some watch a movie multiple times for scenery, character portrayals, to understand a situation or predicament, feel good, feel sad or many other reasons. Some rewatch a movie with hopes the story will end differently. The subsequent viewings are not what I reference. You want to be fully aware of yourself and open to the experience and opportunity to learn something new about yourself or the world and potentially make a difference in your life or in the life of another. If you can hold yourself open, aware and present, I can almost guarantee you an event of miraculous proportions and I remind you miracles can and do happen in a variety of shapes and sizes.

Alternatively, if you think your life is to see how far financially ahead of others you can get, no matter how far ahead of them you are, you are losing. You haven't lost, but it may not look good. Pay attention to what you will give up, push to the side or compromise for more money or to win by your rules. It is always easier to look at others and see what they have done for more money. Try each of those examples on yourself to understand where you have been. Only in looking at and fully understanding the past can you take inventory and forge a path of betterment for yourself and hence the planet.

If you think you are here to stockpile anything and everything, you may think you know what I am going to say, but you probably only know half of it. The half you think you know is my thoughts if you are stockpiling material possessions. Again, you are losing, doubly so if others are denied as a result of your stockpiling.

If you need to push another out of the way to get ahead, you are violating a couple of the laws of nature which include harmony and coexistence.

If you think you are here to harm or kill animals, I assure you are not unless you want to take them on with
rs, indeed you are actually less than for not being able to stand strong on your own. If you think you need to push another down so you can rise, you may consider another path forward with others cheering you on to stand strong and rooting for your success.

If you strive to deny others of anything, you are working on a lesson. Hold on because it may not be pleasant.

Q: What do you mean by a lesson?

I do not mean a lesson in the traditional sense of being in school and learning by someone standing in front of the class and lecturing. I mean more of a life lesson where you learn something about yourself or how you interact with others and the impact that it not only has on you but also on the others around you. You are here to learn about yourself and a lot of that learning comes from interactions with others. When you resist the learning or deny it, you may get a lesson which may not always be pleasant, but it usually is blunt and direct. Choose not to get it or ignore it, and it will come back stronger. These are not my rules per se, it is just the way it works.

Q: There is an unprecedented amount of change in the world at this time and also with nature and individuals. Is all of the change supposed to be happening at the same time?

This is a delicate and complicated question.

As institutions may fall, you will come to the realization that much is not what you think it is.

The world as you have known it will slowly fall away so that a new world will arise. Quite possibly many elements of the new world have been here for a while, but no one has been able to even consider it could be true.

Everything happens for a reason. All of the change that is happening is happening for a reason and there must be a reason it is all happening at the same time.

The world coming together and working to reduce emissions etc was definitely a step in the right direction, however, the world may not have been ready. It must be time to work to understand how the world gets ready other than go backwards and do more damage to the environment and its inhabitants.

In times when changes are occurring, there are often messengers who deliver news etc, both good and bad. Many may not know the words they deliver are messages but in hindsight they will be come to be known as messages, from messengers.

Q: So it is the climate changing or is the climate changing a by product of something else occurring?

It is a fine line and really doesn't matter, it is happening, it is real and it is devastating.

Climate change is largely a result of activity on the planet, the activity of human beings and their disregard for all of its inhabitants. When we consider climate change is a result of a human activity, we may recompute the true cost of a gallon of gas. The impact on the environment of every gallon of gas is far greater than the price paid at the pump. It is far greater by probably 8 to 14 times what is paid to put it in the tank. If the true cost were paid at the pump, there would probably not be the issues there are today with the environment, but once again greed is at work.

Q: Some religions discuss the creation of the planet as part of their religion, for example God created the earth and all on it in a matter of days. Where is the overlap between you and religion?

It is a complex relationship, we will have to delve more into these relationships another time.

Q: Why do you believe we are harming the planet and so many are not concerned about how much they harm the planet or how damaged the planet is?

For so many years, the planet sustained, thrived, recovered regardless of what happened to it. Just like with most everything, the planet has a breaking point. A point at which the strain begins to change the patterns and consistency of the planet and its many ecosystems. Part of it is the planet being over occupancy and the occupants wanting more and more than ever before. If one has more, another wants more and then one wants more than another has and then another wants more than one and so it continues. It is part of a collective mindset wherein a thought appears or is engaged in by one or many and many more join in without processing the thought, its impact, its reality or consequences. Stand on a corner and look up at the sky, even point and see how many will join in having no idea what you are looking at or pointing to. That is an oversimplified example. The danger comes with the thoughts and the collective mindset. Some of the dangerous thoughts are 'the waste I commit is inconsequential' or 'no one knows how much I waste so I don't need to be responsible or accountable for it' or 'the change of the planet is natural' or 'the change of the planet will not impact me' or 'climate change is not really happening.' These are dangerous thoughts whether they be engaged in alone or collectively. It is moments like this that I wish I knew how to swear. Thoughts have impact just as words do.

It is the same with consumption, and related use of resources. One or many go to the store or the mall or order on line and have multiple packages delivered to their door so it may appear as though that is what one should do. So more and more use more and more resources, acquire more items etc etc until we are at a point that the planet is depleted and having a difficult time.

Q: Why do you think we have arrived at this point? How did we get here?

First, I have been very generous with all of you.

A lot of it is ego. Most who acquire money and significant amounts of money do little to help those with less. There is always someone with more and they will strive to get more and more as if they want to get to the next level. Reminds me of the huge houses built to house a relatively few people and the vast unused spaces in them.

The other culprit is the desire to do it all for the money. I don't have any desire or need for money, it is an unknown in my world other than to watch the power it has to corrupt and destroy so many. Destruction of the planet and even polluting the atmosphere and immediate space around the planet is in effect being done for money. It is done to save someone money at the expense of the planet. The planet cannot be bought nor can regeneration of the planet be accomplished only with money. Too many are looking the other way and letting it occur.

Q: What would help people understand their impact on the planet and their next level of responsibility for those impacts?

First, let's have everyone understand they are a guest, not an owner. Anyone who thinks they will own anything forever is a bit out of touch with reality.

After we have the abstract conversation, the reality is there will probably be a disappearance of resources on the planet. It seems many do not understand their impact until it is so impossible for them to miss the impact that then they can understand the impact. When there is no water or they have to stay inside because the air is too polluted to be outside, they may understand the impact, but they will probably still not understand or take ownership of it. You really must understand that you are responsible, no one else, and yes everyone else as well.

*Q:	Is climate change the biggest threat to mankind and the rest of the planet?*

Of course it is and you and I would not be having this conversation if there was not an imminent, significant and irreversible danger to the planet and each of its inhabitants. The imminent nature of the threat is truly about to happen, actually it is happening. So many believe it is reversible, but 100% of the damage cannot be nor will it be reversed. It is not possible to bring a lost species back to life.

Those who look the other way or hold the false belief that the threat is not immediate are not only trying to fool themselves but they are also placing countless others, actually everyone, in jeopardy. My future is not a political issue but many want to make it one. I am not a political being, and as long as I am able, I will protect all to the extent I can. To those who want to play games with the planet, ignore the health of the planet and not help to save the planet, I remind you of the golden rule and its variants. They may think am not real until they are made to be a believer.

Q: That almost sounds like a threat, is it?

I am not making nor would I make a threat ever. Make no mistake though, I will not idly sit by and let a relatively few put so many in jeopardy and effectively initiate and accelerate the demise of an entire planet. Those who are flat out lying or downplaying the plight of the planet for political or monetary gain may want to reevaluate their motives and intentions and subsequently reconsider their 'truth.' There can only be one truth at any one time on a matter. You can weigh either 180 or 200lbs, you cannot weigh both at the same time.

Q: The news just keeps getting worse and worse. There is weather or catastrophe happening almost daily. This is not how the world or the weather was ten or even five years ago. Is this the natural progression of the planet?

Natural can be a complex and often misunderstood word. As we have discussed, everything natural is not always good and wholesome. The word natural describes something existing in or formed by nature or based on things in nature. Technically, everything on this planet is based on or formed by nature. Natural means no more than that. The word has been used to imply it is pure or beneficial to a person, which is not always the case.

Back to your question, it is not the natural progression for the planet to find itself completely out of balance and unable to regulate itself just as it would not the be the natural progression for an individual to find itself completely out of balance and unable to regulate itself. If one of you were to find yourself in this predicament, you would know you were very ill. I am going to be blunt again and tell you the planet is not in good health, in fact, it does not look good.

If you were as sick as the planet, you would most definitely be asking for help, and expecting it. In case I have not been clear, I need help.

Q: What are some of the worst events or plights happening to the planet at this time?

There is not one or a few biggest violations, there are many.

Forest fires not only elevate the heat on the planet at the time they are burning, but they destroy environment and habitat for an extended period of time. The environment may or may not be able to restore itself over a period of time. If it does, it will never be the same. If it does not, it will probably be mostly barren and remain so until another event allows it to regenerate. Forest fires are more destructive that one can imagine.

Water contamination with pesticides and fertilizers, run off after forest fires, plastic in the oceans and flood pollution are devastating. Water holds so much life and life force that is essential to most every living being on the planet. The pesticides and fertilizers from the land stay in the water for an extended period of time and then in bodies. Plastics in the oceans and most all bodies of water are literally changing the chemistry and vital organs of so many beings, it is frightening to me. So many impacting so many.

The quest for and production of oil and gas is difficult on the planet and its inhabitants. Some day it will come to a logical and necessary conclusion. It is hard for many to see and end is possible, but consider there may be an alternative that is unknown to date.

It could be possible to clean up a lot of this, but it may not be considered to be cost effective. There is a value for good health that cannot be quantified in dollars and sense.

It is time to learn and verify. People need to educate themselves as to what the issues are and explore solutions. Some of the most amazing solutions in the past have come from those who were pondering a problem or who looked at a problem from an alternative or personal perspective. Think about your future, you don't need to be a psychic but you do need to look past tomorrow or next weekend. Everyone needs to stand back and take a look and put the pieces together.

Again, you are not here to serve and feed your ego, you are here to love and experience love. Yes, when I use the original L word, it may sound like what many of you may refer to as airy fairy. It may have been airy fairy in days past, especially when the journey was to pursue the dollar at any cost. It may have been airy fairy in days past, when old white men were unequivocally in charge, but times are changing and evolving. It is time to really experience relationships with others, learn about yourself and understand how you influence and impact others on their journey on the planet. In doing so, you will come to understand love in a new light.

Q: It seems like as far back as records of history have been kept there have been wars, are wars a part of the natural progression of humans on this planet?

This is a convoluted question as the more we look at it the more there is to look at. Wars start as disagreements or more than one wanting to own or control something or someone. This view of war is fairly simple. Most claim to not want war, which they believe is the right thing to say. It seems like only in the last fifty years or so has the actual cost of war been acknowledged. The toll on those who go to war are horrendous. There is no justification for destroying the lives of those who are sent to war as well as those who love them. I don't think there has ever been a war with a winner. One side may believe they won but the actual cost to win also makes them a loser. The cost on the planet is gargantuan. As much as wars are romanticized, they are never worth it. Any people or nation wanting war should reconsider their desires.

If the progression of humans requires them to engage in wars, possibly you can declare all the necessary wars have occurred. Wars are another example of failed leadership.

The new war many are calling for is a war with the planet and the changing of the planet. It is now safety and survival in every day life, as opposed to the wars and aggressions from days gone by that is the concern. Threats include rising tides, higher temperatures and stronger and unpredictable storms.

Q: What about space travel?

The idea of having people leave the planet is fine with me. Some should consider leaving the planet and not looking back, but I will not name names. I think I have stated before the planet is above occupancy. The idea of growth again, growth is not about producing and selling more of anything. Growth is not about having more and more without reason. Of course, there will always be someone who will want more and at some point in their time here, they will realize that was not a good idea. There is always the realization, and I am sympathetic to the fact that everyone looks at it differently. There have been advances to help people live on the planet in a more harmonious manner with me and one of them is not to burn a lot of fossil fuels to almost leave the planet, it is ridiculous.

Q: But what about making toxic items off the planet?

I agree toxic items should be banned from the planet. Any fool who thinks surrounding your home with toxins is just that. Explain to me how toxic production can be moved off the planet and not impact the planet. The next issue will be the ozone and damage to it. The planet will not survive without balance. There needs to be a balance between the life on the planet and the sun and energy coming in from space. There is very little understood about space on this planet, it is intentional. I assure you there is nowhere to go that is close to what there is here, but there is always someone who wants to say there is. Going after it will cause horrific damage to the planet. Consider this a warning. The fossil fuels that will need to be burned to get to space on a large magnitude will almost certainly drain the planet of life as you know it.

Q: *What can we do to help?*

First and foremost, stop asking what 'we' can do. Ask 'what can I do'. There may be a we, but there is no individual accountability in we. When most people hear we, they immediately assign duties to others so they don't need to act. People have no problem with saying 'we' need to do something. People understand using we means they are off the hook. Like when a child says 'we need to go to the store for ice cream', they understand they will not drive nor pay for the ice cream. If you want to use the word we, add it on to a declaration of what you will accomplish. If there is one thing you can do to help the planet, it is to get responsible about your use of the word 'we' and how it is an opportunity to avoid doing anything of significance. Use the word 'we' when you are referring to a massive number of 'I's who are taking action and making a difference in the collective future.

Q: *What do you want to create for the world in the next years?*

I would like to have humans do what it will take to have the world be more sustainable, actually I should say have the planet survive. The planet will survive, there may not be a lot of humans on the planet by the time the humans are done with it, but the planet will be here.

Q: *Are you capable of making adjustments to have life continue?*

I adapt every day to survive. Many species are more capable of adapting than others at adapting. Unfortunately, one of the species that is slowest to adapt is humans. It takes generation after generation to adapt and the amount of time available is minimal. I doubt humans can morph to survive the elevated temperatures, etc quick enough. Humans can however make adjustments and take steps to help the planet sustain and eventually recover and thrive again. Only by working for the planet will humans accomplish their survival.

Q: What is our cheat sheet of essential to dos?

Stay present/
Feel the wind in your hair and on your eyes.
Experience the touch of everything.
Feel the energy and presence of another and let them feel your energy and presence.
It really is all a miracle.

Laugh often and include yourself in the joke.

Stand up for me, have me feel the love.

If you are living only for money, you are missing out on all of the things that money cannot buy and that are not for sale. Luxury fades, nature hopefully is forever.

Help me in everyway possible. We have a synergistic relationship and need to mindful.

Q: What final thoughts would you like to leave us with?

Your time here is finite. There is no more for me to say about that other than to say it again, your time here is finite. Enjoy love while you are here and create love while you are here. You will create something while you are here, it may as well be love. I love you and I always have. You should feel that in your body as your read this. Get present and stay present. There has been a lot placed on this planet for you, enjoy. Don't forget to be kind to others, yourself and the planet. ❤

An interview with blunt, loving and hopeful moments. Mother Nature has messages we all need to hear and take to heart to understand where the planet is and what we can do as occupants. Experts tell us the planet may be on borrowed time. The interview is direct with its information and messaging with some lighthearted insights. Every page has insights for reflection as well as conversation and debate with others.

Bio

Like most of us, Oscar Shick has a unique, personal and strong connection with and attachment to nature. He grew up amongst trees and fields under a blue sky. He watched the trees disappear and the fields fill with litter while the sky dimmed. An intuitive, Oscar began to receive communications and figured out it was her, Mother Nature, communicating and asking for assistance in getting her message out. He asked questions and her answers were forthcoming. 'She is excited to be in communication with us and wants to work together to restore the planet.'

Oscar has spent his time on the planet interacting with and paying attention to fellow humans, their thoughts, actions and collective progression as a species. Some years have been better than others for humans and the planet just as some days are better than other days. Oscar is committed to elevating the collective consciousness, awareness and the resulting happiness of all living beings and he is committed to continue to do his part to get us there. Oscar will do his part quietly and as an anonymous occupant on the planet using his pen name.